INTRODUCTION TO
PHYSICAL SCIENCE

INTRODUCTION TO PHYSICAL SCIENCE

A SCIENCE CONNECTED LAB MANUAL

Science Connected

Introduction to Physical Science: A Science Connected Lab Manual

Published by Science Connected, Inc.

Writers: Emily Rhode, Steven Spence, Dr. Jonathan Trinastic

Editor: Kate Stone

Lead copy editor: Jess Romaine

Additional copy editors: Rachel McCabe, Sabrina LeRoe

Cover photographer: Jillian Wheeler

Cover designer: Raychelle Anne Ignacio, Marketplace Designers

© 2018 by Science Connected

180 Steuart St. #190213, San Francisco, CA 94105 USA

www.scienceconnected.org

Contents

How to Use This Book

Each of the sections below will appear in each experiment. Read the short descriptions to understand what type of information is included in each section before moving on to individual experiments.

The contents of this guide are aligned with the Next Generation Science Standards (NGSS).

What It's About

In this section, you will learn about the basic scientific premise of the experiment. You will find out what the student will do, what the student will learn, and why this experiment is interesting. Read this section with your student to help them understand what the experiment will be about.

What You Need

Here you will find the materials needed to successfully complete the experiment. If you do not have a specific material, you can often make substitutions without changing the procedure or outcome of the experiment.

Useful Words

Here you will find:

- What new words students will learn from this experiment

- What words they will need to know to talk about the experiment

Be sure to go over these words and their definitions before you begin the experiment. You can start by saying one word and asking your learners if they know what it means. After you go over the definitions, it might be helpful to see if your student can use the words in a sentence in the correct context before starting the experiment.

What to Do

Here you will find the steps you and your student will take to complete the experiment. Be sure to take your time and follow the steps in order. Have the student record their observations along the way, and ask them questions such as the following: What do you think will happen? What did you notice? Why do you think that happened?

If something does not work, talk about what happened. Just because the experiment did not work does not mean that it was a failure! Feel free to start over and try again, and make sure to compare your results from the second try to the results from the first try. Science is all about making mistakes and learning from them.

Science to Know

Head to this section for more information about the science behind each experiment. This section is meant to help parents and teachers further explain the scientific concepts. We provide more details about the science, and we take a closer look at what might have happened in the experiment.

Applications

Each chapter in this book includes a corresponding story about how the science behind the experiment is being used in research and industry around the world. We invite you to read each story before or after conducting the experiment. You and your learners can connect concepts used during the labs in this book to recent scientific research, discover how scientific experiments are used in a broader context to develop new technology, and develop a more robust understanding of the world in which we live.

Demonstrating the Forces of Flight

NGSS Standards: 3-PS2-1 Forces and Interactions

What It's About

Have you ever wondered how birds are able to fly? Or how an airplane that weighs so much can make it off the ground? The air around us has lots of properties, including mass. Air pressure and gravity can make flying a tricky task. There are many forces that make it possible for animals and airplanes to fly. In this experiment, you will demonstrate one of these forces by simulating the flow of air over an airplane or bird wing. You might be surprised by the results!

What You Need

- A pencil

- A piece of notebook or printer paper

- Scissors

- Tape

- Two pieces of string, each about 30.5 cm (12 inches) long

- Two latex balloons

- One ruler or wooden dowel at least 30.5 cm (12 inches) long

Useful Words

- Air pressure: the force exerted on a surface by the weight of the air

- Lift: a force created by differences in air pressure that directly opposes the weight of an object

What to Do

1. Use scissors to cut a strip of paper from the long side of the piece of paper.

2. Tape one short end of the strip of paper to the middle of the pencil.

3. Hold the pencil in both hands and bring the end of the strip of paper that is attached to the pencil up to your lips.

4. Blow a steady stream of air over the top of the strip of paper. What happens to the strip of paper? Did you expect this to happen? Why do you think this happened? What does this have to do with air pressure and lift?

5. Now blow up each balloon to approximately the same size.

6. Tie a piece of string to each balloon.

7. Tie the other end of each piece of string to the ruler so that the balloons are hanging from the ruler at the same length and not touching each other.

8. Hold the ruler up so the balloons are hanging in front of your face.

9. Gently blow a steady stream of air through the space between the balloons. What happens to the balloons? Did you expect this to happen? Why do you think this happened? What does this have to do with air pressure and lift?

Science to Know

The wing of an airplane is shaped kind of like a teardrop. The same amount of air flows above and below the wing as a plane flies. Because of the teardrop shape, the air that flows over the top of the wing has to travel a longer distance than the air that flows under the wing. This means that the air on top of the wing has to travel faster than the air below. The faster air molecules spread out and are less dense than the air molecule below the wing. Higher density below the wing and lower density above the wing creates an unequal force and the plane rises.

When you blow over the top of a strip of paper, you are making the air molecules move quickly, which is making them spread out. This creates an uneven force, or lift. The same thing happens when you blow air through the space between the balloons. Lift is one of the forces that helps planes overcome the force of gravity during takeoff, and keeps planes in the air.

Application: How Planes Fly

Do you know any airline pilots? Piloting an aircraft is very different from driving a car, riding a bike, or other forms of ground transportation.

Yaw, Pitch, and Roll

There are three axes in play during flight. First, the airplane can turn left and right. This is called yaw. Second, the airplane's nose can move up and down. This is called pitch Third, the body of the airplane can lean left and right. This is called roll.

Why Do Most Accidents Occur During Takeoff and Landing?

The speed of an aircraft is not measured by how fast it's rubber wheels turn on the pavement, but instead by air that rushes past the plane and is captured in a tube. Measuring the air pressure inside that tube reveals the speed of the aircraft. These factors are crucial to safety because most aircraft accidents occur during takes offs and landings. Why? Having sufficient airspeed to maintain lift is what you need to not stall out-- a situation that obviously cannot be remedied by pulling over and checking the manual or lifting the hood.

Airspeed, Lift, and Drag

By the design of the wings (more rounded on top, flatter on the underside) the air flow is broken up. The air passing over the top of the wing is thinner than the air traveling under the wing, allowing the aircraft to rise. This is known as lift. Lift can be lost at high altitudes where the air is thinner, or in hot climates where the air expands and is less dense. Wherever there is less air, there is less lift for the plane in flight. The pilot must rev the engine at the proper time to not stall, lose air speed, and succumb to the pull of gravity. That information is provided to the pilot via the instrument panel which will also give an audible warning when airspeed is insufficient to maintain lift.

Make It Move: Measuring the Static Friction of a Shoe

NGSS Standards: 3-PS2-3 and 3-PS2-4 Forces and Interactions

What It's About

Friction is a force. When something is not moving, we say it has static friction. This means that it will take a certain amount of force to make an object slide past the object it is resting on and start moving. Some things have more static friction than others. Thus, walking on ice feels more slippery than walking on carpet. In this experiment, you will find out how much force you need to overcome static friction and make something move. To test this, you will pull a shoe over different surfaces.

What You Need

- A shoe without a heel (any size is fine)

- A rubber band

- A ruler or measuring tape

- Weights to place inside the shoe (coins, rocks, etc.)

- Cooking oil

Useful Words

- Friction: the resistance that one surface or object encounters when moving over another

- Force: something that causes a change in the motion of an object

- Static friction: force between two or more solid objects that are not moving relative to each other

What to Do

1. Cut a rubber band loop so that it makes one long rubber band.

2. Tie one end of the rubber band to one end of a shoe.

3. Place the shoe on a smooth surface like a table or counter top.

4. Hold a ruler alongside the rubber band so that the end of the ruler lines up with the place where the rubber band is tied to the shoe.

5. Pull on the loose end of the rubber band until the shoe starts moving.

6. Measure how far the rubber band had to stretch before the shoe moved. Record your measurement. Was it easy or hard to move the shoe?

7. Add weighted materials (such as rocks or coins) to the shoe. What do you think will happen when you try to pull the shoe now?

8. Pull on the loose end of the rubber band until the shoe starts moving.

9. Measure how far the rubber band had to stretch before the shoe moved. Record your measurement. Did the rubber band stretch more, or less than when the shoe had no weight in it?

10. Tape a piece aluminum foil to the ground.

11. Place the shoe on the aluminum foil.

12. Pull on the loose end of the rubber band until the shoe starts moving.

13. Measure how far the shoe had to stretch before the shoe moved. Record your measurement.

14. Now try adding a small tablespoon of cooking oil to the aluminum foil. Spread the oil into a thin layer on the foil.

15. Pull on the loose end of the rubber band until the shoe starts moving.

16. Measure how far the rubber band had to stretch before the shoe moved. Record your measurement. Was there more static friction between the shoe and the aluminum foil with cooking oil or without it?

Science to Know

Friction affects our lives every day. When you go down the slide at the playground, friction eventually slows you down at the end of the slide. Friction makes it hard for you to push a heavy piece of furniture across the floor. And friction helps to slow your car down when the driver uses the brakes. Friction is a force that acts in the opposite direction of the way something wants to move.

Static friction is the force between two things that keeps them from slipping or sliding past each other. When you want to run down the sidewalk, the static friction between your shoe and the sidewalk lets you move forward. If there were no static friction, you would just be running in place as if you were on a treadmill.

Kinetic friction is the force that slows down an object that is already in motion. This force will try to reduce the speed of the sliding until the object comes to a stop. Without kinetic friction, a soccer ball would keep rolling forever after you kicked it (if no objects were in the way to stop it).

Application: New Rubber Grips Icy Surfaces

Winter storms dumped record amounts of snow on the East Coast and other regions of the United States in 2015, forcing many people to navigate icy sidewalks and roads. However, treacherous travel by foot may soon be a thing of the past thanks to a team of researchers from the Toronto Rehabilitation Institute and the University of Toronto, Canada, who are working on a new rubber sole to help pedestrians get a better grip on slippery surfaces.

The material is made up of glass fibers embedded in rubber, and it could be used to make slip-resistant rubber soles for winter boots.

Ice can be an especially hazardous surface when the temperature gets close to zero degrees Celsius and a thin layer of liquid water forms on top. Cleated footwear provides effective winter traction because the cleats dig into the still solid ice beneath the slippery layer of water, but because they are slippery on other surfaces, they can be dangerous if the wearer does not take them off before going indoors. In response to the need for better winter footwear, the research team developed a new method to manufacture rubber soles that dig into an icy surface on a smaller scale than cleats do.

"I think anyone who has slipped or fallen on ice can testify that it is a painful experience," said Reza Rizvi from the Toronto Rehabilitation Institute. "Now imagine being frail or disabled - a slippery sidewalk or a driveway is all that it takes to trigger a life-changing fall. A serious fall on ice resulting in a hip fracture can be a death sentence for an older adult."

Tilak Dutta, also from the Toronto Rehabilitation Institute, points out that falls are only part of the problem. "Equally important are the many older adults who feel trapped indoors for long stretches in the winter because of the fear of falling. The lack of activity and isolation have major negative impacts on health. We need to give older adults better footwear so they feel confident maintaining their outdoor activity levels in the winter."

Versions of the new rubber sole are being tested in a self-contained lab that can be tilted to create sloped floors that are covered in ice and snow. The incline is increased until the volunteers who are testing the special shoes start to slip. The lab is also equipped with a padded wall and a safety harness so the volunteers don't get hurt.

A Cross Between a Rubber Sole and Sandpaper

The material is made up of thermoplastic polyurethane, a rubbery plastic, embedded with tens of thousands of tiny glass fibers that protrude out of the rubber like microscopic studs and give the material the feel of fine sandpaper.

The material looks like regular rubber and will stretch and deform in similar ways. The material also performs just as well as regular rubber on hard, dry floors. On ice, however the rubber-glass fiber composite provides significantly better traction. The researchers hope that their work will prevent many slips and falls on winter ice.

Existing methods for fabricating the material require first extruding a rubber slab with glass fibers running parallel with the surface. The slab is then cut and reoriented so that the fibers stick out of the surface like tiny pins.

"The materials required for creating a high friction composite are not expensive, but the process of slicing and rearranging the rubber is not easily scalable," Rizvi says. The team has been working on ways to automate the process so that the material can be cheaply mass-produced and be made to last longer. The prototype rubber sole quickly loses its non-slip qualities with use, so it will not be appropriate for commercial footwear for walkers and hikers until its robustness is improved.

So, what excites the research team the most? "I am most excited about taking my research and having it applied to a serious societal issue of winter safety," Rizvi says.

"This work has the potential to have a real impact on the massive, expensive public health problem of winter falls, not to mention the dramatic improvement in quality of life for those living in northern climates," Dutta adds.

This research is published in the journal *Applied Physics Letters*.

How Do Different Materials Affect Temperature?

NGSS Standards: K-PS3-1 Energy

What It's About

When the sun shines on something, the object heats it up. Have you ever noticed that different colored objects sitting in the sun feel warmer or cooler when you touch them? What does it feel like when you wear a dark shirt on a bright summer day? In this experiment, you will test to see which color paper will absorb heat energy better.

What You Need

- One piece of white construction paper

- One piece of black construction paper

- Tape

- A lamp with an incandescent bulb (or a heat lamp)

- A thermometer (digital is fine)

- Something to keep time

Useful Words

- Absorb: take in or soak up by chemical or physical action

- Heat: a form of energy associated with the movement of atoms and molecules in any material

- Reflect: to move in one direction, hit a surface, and then quickly move in a different and usually opposite direction

- Temperature: the degree or intensity of heat present in something

What to Do

1. Fold each piece of paper in half.

2. Tape two edges of each piece closed. Be sure to leave one edge open so the paper forms a pocket.

3. If using an analog (not digital) thermometer:

    a. Place the thermometer inside of the white envelope.

    b. Turn on the lamp.

c. Place the envelope with the thermometer inside of it under the lamp.

d. Wait 5 minutes.

e. When 5 minutes are up, read the temperature of the thermometer and record it.

4. If using a digital thermometer:

a. Turn on the lamp.

b. Place the empty white envelope under the lamp.

c. Wait 5 minutes.

d. When 5 minutes are up, turn on the thermometer and place it inside of the envelope.

e. When the thermometer beeps, read and record the temperature.

5. Repeat the same procedure with the dark envelope. Which envelope had a higher temperature? Why do you think that is? You can repeat the same procedure with different colored paper. How do you think the color will affect the temperature inside the envelope? What might happen to the temperature if you used a different material than paper? Give it a try!

Science to Know

Visible light is a type of electromagnetic radiation. These waves can be seen, but there are many types of electromagnetic waves that cannot be seen by human eyes. Infrared waves are one type of electromagnetic wave that cannot be seen by our eyes. The sun heats up objects by moving energy in the form of infrared waves.

When an object soaks up, or absorbs, the infrared waves, its temperature goes up. Some materials absorb heat energy better than others. Some materials reflect heat energy. Dark surfaces such as roads and rooftops absorb more infrared radiation than lighter surfaces including trees, grass, or water.

Application: Hot Towns, Urban Heat Islands

Have you ever noticed on weather reports that cities seem to be hotter than the surrounding areas? That's a result of the Urban Heat Island (UHI) phenomenon. According to the Environmental Protection Agency, urban areas with 1 million or more residents have a mean annual temperature 1°C to 3°C warmer than their surroundings. At night, the effect is even more pronounced, with city temperatures reaching up to 12°C hotter. With more than half (54 percent) of the world's population living in urban areas and the trend toward urbanization increasing, UHIs have a significant effect on the way populations experience climate.

How Is the Urban Climate Different?

Urban environments have a high percentage of surface area covered by buildings, streets, and parking lots. These surfaces are generally dark, and water tends to run off quickly. The dark surfaces absorb more infrared (heat) radiation from the sun than lighter rural areas do. As the day goes on, these surfaces radiate the absorbed heat, increasing air temperature well into the night. Also, roofs and paved areas are drier than the soil in rural areas, which limits a cooling effect from water evaporating.

Impacts of Urban Heat Islands

Positive effects of UHIs include a reduced need for heat in winter and a longer plant-growing season. Negative effects include increased energy consumption, higher emissions of air pollutants releasing more greenhouse gases), and impacts on human health. We'll look at these effects in the following sections.

Energy Consumption and Air Quality

Electrical demand in the United States increases 1.5 percent to 2 percent for every 0.6°C increase in temperature. We know that urban heat islands add 1°C to 3°C to regional temperatures. Thus, a typical urban household uses between 1.5 percent and 6 percent more energy for cooling than a rural household does.

Burning more fossil fuels to generate electricity results in increased emissions of sulfur dioxide, nitrogen oxides, carbon monoxide, carbon dioxide, and fine particulate matter. It also contributes to increased ground-level ozone formation and smog.

Health Effects

Prolonged high temperatures cause a range of human health problems. They make existing medical conditions worse and can cause rashes, cramps, heat exhaustion, heat stroke, and even death. During extreme heat events, mortality doubles for every 1°C increase in temperature. In hot conditions, an extra 1°C to 3°C degrees due to an UHI can be lethal.

According to the Center for Disease Control (CDC), during extreme heat events—such as the one in Chicago in 1995 that resulted in 650 deaths—infants, the elderly, and homeless or low-income people are most at risk. Heat was the number-one cause of weather-related deaths (approximately 7,800) in the United States from 1999 to 2009.

In countries where air conditioning is not as widespread, extreme heat waves may kill even more. The 2003 heat wave in Europe killed as many as 30,000 people. There were 14,800 deaths that year in France alone [4]. Satellite measurements (LANDSAT) confirmed that nighttime temperatures in Paris were 4°C higher than temperatures outside Paris at the time. Multiple hot days and hot nights in a row are deadly because vulnerable people cannot recover adequately from the heat.

Reducing Effects of Urban Heat Islands

Absorption of infrared (IR) radiation by covered surfaces in cities is the key factor in producing urban heat islands. One way to combat this effect is to choose materials with a high reflectivity (albedo). Urban areas offer three large opportunities for adding more reflective surfaces: An analysis of Sacramento showed that 28 percent of the city area consists of rooftops, 16 percent of streets, and 14 percent of parking lots and driveways. This means that 58 percent of Sacramento's surface area absorbs heat during the day and releases it slowly at night. Chicago, Houston, Sacramento, Baton Rouge, and Salt Lake City were also analyzed and had similar results. Scientists estimate that these cities have the potential to modify the urban albedo over 18 percent of the surface area.

Smarter Use of Building Materials

Using white surface coatings for roofs and walls would increase reflectivity from 5 to 10 percent for lighter roofs and as much as 50 to 80 percent for darker roofs. Reflecting more infrared energy away from the building keeps temperatures lower. In fact, the temperature difference between a dark roof and a reflective roof may be as much as 33°C.

Pavements and parking lots can be constructed with light-colored materials. Mixing light-colored stone aggregates into asphalt or switching to concrete improves the reflection of infrared radiation. Some of this reflected radiation will strike nearby buildings, which may require reflective glass windows or additional insulation.

Urban Heat Islands: Trees to the Rescue

Another option for cooling roofs is to cover them with trees and other types of vegetation. Chicago's City Hall is a well-documented example. Green roofs are more complex to build and maintain than roofs simply coated with highly reflective surfaces, but the vegetation also improves overall air quality and reduces runoff.

A study in Mexico City concluded that temperatures could be reduced 1°C by planting trees. Planting either 63 river red gum trees (Eucalyptus camaldulensis) or 24 American sweet gum trees (Liquidambar styraciflua) per hectare (10,000 square meters) achieves this cooling effect. Shade for specific buildings can be further optimized by planting trees or vines on the east or west sides, blocking low-angle rays that contribute to heating buildings in the morning or afternoon.

The type of tree canopy and the suitability of the growing climate are important factors in deciding which trees to plant. Trees provide a simple and cost-effective method to reduce UHI effects. Trees that lose their leaves (deciduous trees) also have the advantage of not blocking the sun during winter. This preserves the winter warming effect and reduces energy used for heating.

Harnessing the Sun

Another option for controlling UHIs is using solar power panels, which also reduce harmful emissions from fossil fuels. Thermal panels reduce heating costs, while photovoltaic panels generate electricity. Solar panels also cool buildings by shading them from infrared radiation.

Researchers estimated that solar panels would result in 3 percent more wintertime energy usage in Paris. A 12 percent summertime reduction in air conditioning would more than offset the winter increase. Additionally, solar panels would reduce the nighttime UHI effect by 0.3°C.

A research group simulated large-scale deployment of solar panels. They found that global temperatures would decrease by an average 0.04°C. Their simulations also uncovered possible changes in regional precipitation. Nevertheless, the researchers concluded that using solar panels is better than using more fossil fuels. Burning fossil fuels is projected to increase global temperatures 1°C to 3°C by 2100. The results of the study will help policy makers reduce fossil fuel emissions with minimal side effects on precipitation.

A Real-Life Game of SimCity™

Policy makers and city planners have multiple options to combat UHI. Roofs account for up to 40 percent of UHI effects. They can easily be improved with solar-reflective materials, green roofs, or solar panels. Solar panels reduce both UHI heat effects and fossil fuel emissions. Vegetation, especially trees, reduces the summertime UHI heat effects through a combination of shade and evaporation. Vegetation also has a positive effect on improving air quality and reducing runoff.

How Are Colors Created?

NGSS Standards: 1-PS4-3 Waves and Their Applications in Technologies for Information Transfer

What It's About

Color is all around us—in the green trees, the blue sky, yellow flowers, and so much more! Even though there are only three primary colors (red, yellow, and blue), the human eye can see millions of colors (about 10 million to be exact!). We are most sensitive to the colors red, green, and blue. Scientists are always learning more about how our eyes detect these colors. In this experiment, you will use colored paper and colored plastic filters to add and subtract colors and create different color combinations!

What You Need

- A dark room

- A flashlight

- Construction paper (green, blue, and red)

- See-through colored cellophane paper or plastic wrap (green, blue, and red)

- A rubber band or tape

Useful Words

- Absorb: take in or soak up by chemical or physical action

- Color: the property of an object producing different sensations on the eye as a result of the way the object reflects or emits light

- Light: visible radiation having wavelengths in the range of 400–700 nanometers

- Reflect: to move in one direction, hit a surface, and then quickly move in a different and usually opposite direction

What to Do

1. Make a data table like the one on the next page.

Paper Filter	White	Green	Blue	Red
None				
Green				
Blue				
Red				

2. In a dark room, turn on the flashlight and point it at the piece of white paper (Paper: White, Filter: None). What color is the paper now? Record the color on your data table. Was this what you expected? Why or why not?

3. Repeat step 2 with the green, blue, and red pieces of paper (Filter: None). Did the color of the light change?

4. Use tape or a rubber band to cover the lens of the flashlight with a green plastic cellophane filter. Shine the flashlight first on the white paper, then on the green, blue, and red papers and record what color each piece of paper appears to be. What did you observe? Did you expect this? Why or why not?

5. Repeat step 4 using the blue filter and then the red filter. After each test, record your observations on the table. Were the colors that you saw what you expected? Why or why not? What do you think was happening to the light? Do you think you could make any other color combinations? Give it a try!

Science to Know

Visible light is made up of energy. When waves of light shine on something, some of the light is absorbed and some is reflected into our eyes. That reflected light is how we see an object.

Depending on the length of the wave of light that is reflected, we see different colors. Our eyes send a message to tell our brain what colors of light have been reflected, and our brain mixes and blends the colors for us. That is how we can see such a wide variety of colors! Blue is one of the shortest wavelengths of light. It has a wavelength of 400 nanometers. Red light has a much longer wavelength of almost 700 nanometers.

When you look at a blue butterfly, you see blue because all the other wavelengths of color have been absorbed by the butterfly and only blue wavelengths (or color) are reflected into your eyes. When you look at green tree, all the other wavelengths of light have been absorbed by the tree except for green. The green light is reflected into your eyes.

In nature, animals often use bright color to attract mates or to scare off predators. Sometimes animals will use colors to camouflage themselves to hide from predators or even prey. Plants use color to attract pollinators or to warn of the danger of eating or touching the plant. These organisms may have very complex ways that they display their colors, and some of their colors can only be seen by using very powerful microscopes.

Application: Structural Coloration in Bird Feathers

If you are familiar with *Pimp My Ride*, a television show about customizing old cars, you will know that one of the key elements in each customization is a bold paint job. It turns out that nature is ahead of television.

Biomimetics is the science of harnessing (or mimicking) nature in the design of human-engineered products. Aircraft wings have been improved with inspiration from birds, bats, and even sharks. Geckos and their amazing gripping properties have inspired the development of adhesives and robotics. Structural coloration, as found in butterfly wings and bird feathers, is also inspiring work in optical computing, thin film optical coatings, and color-changing paints.

What if I Told You a Peacock is Brown?

Peacocks are known for their bright colors, but when their feathers are wet they appear brown. This indicates that the bright colors aren't a result of pigmentation: something else is going on. Indeed, the peacock achieves its stunning plumage display through structural coloration, more commonly known as iridescence.

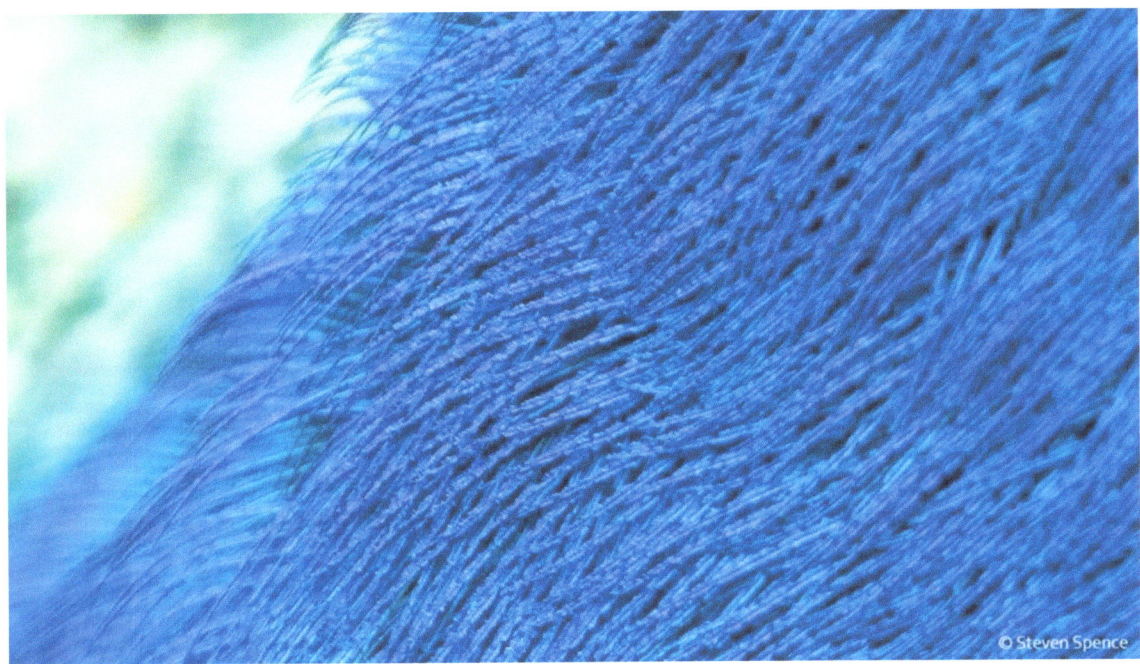

Close-up of a peacock's "blue" feathers

Structural Coloration, Shifting Colors

A surface is said to be iridescent if it appears to shift colors as the angle of illumination changes. Structural coloration results from a light wave interaction with a structure of similar size or half of the size. The light can be reflected, refracted, or sometimes both, and the effect is angle dependent, according to Bragg's law.

Depending on the angle and the wavelength, light waves may constructively or destructively interfere with each other. If waves destructively interfere, they minimize or completely cancel each other out. Constructively interfering waves increase in size and can double in strength (amplitude).

Magic with Mirrors

Thin film interference occurs when light reflects off the top and the bottom of a nano-scale layer. A common example of this is light hitting a film of oil on water. Light is reflected once from the oil on top of the water and a second time from the water-oil interface. Soap bubbles are another example.

A thin layer of oil on a wet asphalt road

Is Color a Wave or a Particle?

We won't come close to settling that question right now. In this case, light is behaving as a wave. Unlike the chameleon, our peacock makes use of 2-D photonic crystals. Photonic crystals are structures that allow transmission of light only at certain frequencies. To function, the structures need to be approximately half the wavelength of the light transmitted.

A nanometer (nm) is one billionth of a meter. The human eye can view colors in the range of 390 nm (edge of ultraviolet) to 700 nm (edge of infrared), thus iridescent violet or blue seen in peacock plumage requires structures in the range of 200-250 nm. For comparison, this is half the size of the structures in a Blu-ray disk (405 nm).

The Secret of a Peacock's Colors

A peacock's feathers have barbs. Each barb has a flat array of flat barbules, structures which disperse the incoming light and create the coloration we see. The peacock's blue, green and yellow barbule structures are nearly square.

Scientists have recently discovered that the brown structures are also caused by photonic crystals of a rectangular shape and not pigments as previously thought.

Color Follows Form: Shining Examples

Besides peacocks many other birds have iridescent plumage-- starlings and ducks are other common examples. Among insects, structural coloration gives some butterflies, flies, wasps, and beetles a bright, metallic sheen.

Iridescent Superb Starling (Lamprotornis superbus), photographed at Vogelpark Walsrode

Applications

Structural coloration has multiple applications in optics and photonics. Distributed Bragg Reflectors (DBRs) have applications in fiber optics, lasers, and optical computing. Thin film layers can be used in anti-reflective coatings on lenses, filters and mirrors. Iridescent paints may be used for cars, but may also provide a new level of intensity and depth to art works.

How Can You Turn Saltwater into Drinking Water?

NGSS Standards: 2-PS1-1 and 2-PS1-2 Structure and Properties of Matter

What It's About

The ability to separate water from solids is important for many basic human needs, including access to clean drinking water. In this experiment, you will compare how well different materials can filter out contaminants from dirty water. You will also explore the different ways to separate materials by using physical and magnetic filters. The principles you learn in this experiment are being used by scientists today to gain access to drinking water from saltwater.

What You Need

- A bowl that can hold at least a liter (a little over 1 quart) of water

- Several drinking glasses

- A measuring cup

- A funnel

- Water (around a liter, or a little over 1 quart)

- Dirt or soil

- A handful of glass marbles

- One coffee filter

- A paper towel

- Aluminum bolts or nails

- Steel bolts or nails

- A magnet

Useful Words

- *Filtration*: a process that separates solids and liquids from one another

- *Magnetism*: a force generated by magnetic fields that occur in some materials. Magnetic objects can be attracted or pushed away from one another.

What to Do

1. Fill a bowl with about a liter (a little over 1 quart) of water. Mix enough soil into the water until it becomes murky and opaque.

2. We will first test how well a very basic physical filter—glass marbles—works to separate out the dirt that has been mixed into the water. Pour the glass marbles into the funnel (be sure that the spout of the funnel is small enough so that marbles cannot fit through!). Set the funnel on top of a drinking glass so that the spout points into the glass.

3. Use the measuring cup to scoop two cups of the dirty water, and pour the water through the marbles in the funnel so that it falls into the glass.

4. Record your observations while comparing the original, dirty water to the water in the glass that has passed through your marble filter. Do you see any noticeable differences? Why or why not?

5. Next, we will test a coffee filter's ability to separate dirt from the water. Set aside the glass with the water that passed through the marble filter. Also, remove the marbles from the funnel. Then, place one coffee filter over the funnel and place the funnel onto a new, empty glass.

6. Pour another two cups of dirty water onto the coffee filter. Again, record observations about the water that filters through into the glass. Is this water dirtier or cleaner compared to the water that passed through the marble filter? Why?

7. Finally, repeat the above two steps using a paper towel as the filter placed over the funnel. Again, record any observations about the water filtered into the glass.

8. Compare your notes about all three filters (marbles, coffee filter, paper towel). What are the major differences between the filters? What differences did you see in the water after filtration? What conclusions can you make about what determines whether a material will act as a good filter to create cleaner water?

9. Not all filters have to operate using a physical barrier! To demonstrate this, mix a handful of aluminum and steel nails or bolts together into a bowl. The goal is to separate this mix into aluminum and steel objects. What's the fastest way to do this other than separating them out one by one with your hands?

10. Try holding a magnet over the bowl. What happens? Can you use the magnet to easily filter these materials? What property of the bolts and nails is being used here to create the filter?

Science to Know

Physical filters work by creating spaces that liquid can pass through but solids cannot. The marble filter should not have worked well to separate the dirt from the water because both the water and the dirt particles could fit through the relatively large holes between marbles. In contrast, the coffee filter has much smaller holes that could trap some dirt, leading to cleaner water passing through into the glass. The paper towel should have performed as the best filter because it only has microscopic holes that can prevent almost all dirt from passing through. Therefore, the key to a physical filter is to leverage the fact that the material to be filtered out is larger than the water molecules.

Not all filters have to operate using physical barriers. A filter only relies on taking advantage of any property that is different between the two materials you want to separate. As an example, the magnetic filter used at the end of this experiment relies on the fact that aluminum nails and bolts are not magnetic and therefore are not attracted to a magnet, whereas steel nails and bolts will latch onto a magnet. We can use this difference in magnetic properties to create an effective filter using the magnet to separate steel and aluminum parts.

Application: Graphene Sieve Turns Saltwater into Drinking Water

Water, water, everywhere,

And all the boards did shrink;

Water, water, everywhere,

Nor any drop to drink.

—*Rime of the Ancient Mariner*

Poet Samuel Taylor Coleridge published those lines in 1798. In 2017, scientists from the University of Manchester developed a graphene-based desalination tool. Soon, more of that abundant seawater might be drinkable after all. This is good news for Coleridge's ancient mariner and for everyone in need of fresh water.

Fresh water is essential. It has been called liquid gold. According to the United Nations, 85 percent of the global population lives in the driest half of the planet, 783 million people do not have access to clean water, and almost 2.5 billion do not have access to adequate sanitation. Additionally, 6 to 8 million people die annually from the consequences of water-related disasters and diseases.

Desalination, or removing salt from seawater to make it drinkable, is not a new idea. According to the International Desalination Association, 18,426 desalination plants were operating worldwide as of 2015, producing 86.8 million cubic meters per day and providing water for 300 million people.

The new technology has the potential to be smaller, faster, and more easily adjustable than existing methods. Preliminary tests of the graphene sieve have been promising. So far, it seems that the sieve can efficiently filter out salts. Next, the effectiveness of the new graphene-based sieve will need to be compared to other desalination membranes already in use.

Speaking of other membranes, yes, there are already several large desalination plants around the world trying to use polymer-based membranes to filter the salt out of seawater, but the process is too inefficient and expensive for widespread use. Thus, finding a way to turn seawater into drinking water more quickly and with minimal expense has been a key goal in the latest research.

Rahul Nair from the University of Manchester is optimistic. "This is the first clear-cut experiment in this regime. We also demonstrate that there are realistic possibilities to scale up the described approach and mass-produce graphene-based membranes with required sieve sizes."

Graphene oxide membranes have long been considered a promising candidate for desalination, but successfully removing salt requires the holes in the graphene oxide membrane the water passes through to be incredibly tiny.

Even though this new desalination technology is restricted to the laboratory for now, in the not-too-distant future it may be used to change an extremely abundant resource—

seawater—into a very rare one—drinkable fresh water. This research has been published in the journal *Nature Nanotechnology*.

How Can Geckos Climb Walls?

NGSS Standards: 2-PS1-1 Structure and Properties of Matter

What It's About

Have you ever wondered how small lizards such as geckos can climb vertical walls without falling off? The answer lies in how tiny particles in the gecko's feet are attracted to the surface of the wall, opposing the force of gravity. In this experiment, you will explore a similar phenomenon known as surface tension that keeps liquids clinging together in much the same way as the gecko's feet stick to the wall.

What You Need

- Two handfuls of pennies

- A glass of water

- Enough water to fill the glass

- A shallow plate

Useful Words

- *Surface tension*: the attractive force between molecules in a liquid that tend to hold the liquid together and resist the effect of other forces acting on it

- *Gravity*: the force that attracts one object with mass to another object. The primary force of gravity we feel is Earth keeping us on the ground. This is because Earth is the most massive object in our environment, so its gravitational force is strongest.

- *Van der Waals forces*: attractive forces between molecules at the microscopic scale

What to Do

1. Fill the glass with water exactly up to its top brim. Then, place the glass onto the middle of the saucer (the saucer will help catch any water that spills).

2. You will now begin adding pennies into the glass, one at a time. It is important to add them carefully to prevent water from splashing over the sides of the glass. Instead of just dropping the pennies into the glass, slide them slowly along the rim and let them gently fall into the water.

3. As you add each penny, observe the behavior of the water at the top of the glass and along the rim. What changes do you see to the water's shape?

4. Continue to add pennies, one at a time, and continue to record observations about changes in the water.

5. After enough pennies have been placed in the glass, the water should eventually pour over the sides of the glass, ending the experiment.

Science to Know

You may have observed that as you added more pennies, the water rose above the brim of the glass without spilling. This effect occurs because of surface tension in the water. Surface tension occurs because individual water molecules are attracted to one another, feeling a force stronger than the force of gravity that tries to pull the water over the sides of the glass and onto the saucer. This cohesive action of surface tension in the water is stronger than gravity when only a small amount of water is above the brim of the glass. When enough pennies are added to the glass, the force of gravity is strong enough to overcome the surface tension in the water and pulls the water over the sides and onto the saucer.

Surface tension doesn't just apply to glasses of water—geckos also take advantage of this attractive force between molecules to stick to walls! These little lizards have tiny fibers called *spatulae* on their toes. The molecules in the spatulae interact with the molecules on the surface of the wall through van der Waals forces. These are naturally attractive forces between molecules that keep the gecko from slipping down the wall. In addition, on certain wet surfaces, the surface tension of the water on the surface helps the gecko keep a firm grasp on the wall.

Application: Why Spiderman Can't Exist, but a Gecko Can

With all due respect to Spiderman, it turns out that physics is against our wall-crawling, web-slinging hero. There is a size limit on who or what can stick to walls: the size of a gecko.

David Labonte and his team at the University of Cambridge Department of Zoology have been wondering why geckos are the largest animals able to scale smooth vertical walls.

Geckos have highly effective and complex foot pads that they use to climb smooth, vertical surfaces. However, anything larger than a gecko would need to have unwieldy large and sticky foot pads. The scientists estimate that to climb a wall like Spiderman does, a human would need to have adhesive pads covering 40 percent of their body surface.

"If a human, for example, wanted to walk up a wall the way a gecko does, we'd need impractically large sticky feet. Our shoes would need to be a European size 145 or a US size 114," says Walter Federle from the Cambridge Department of Zoology.

In different sizes of climbing animals, from tiny mites and spiders to tree frogs and geckos, the percentage of body surface covered by adhesive footpads increases as body size increases. This increasing ratio sets a limit to the size of animal that can scale walls, because larger animals would require impossibly big feet. While such an animal might make for terrific science fiction, such a body would not work in the world as we know it.

Hope for Would-Be Wall Climbers

The researchers are studying the feet of geckos, insects, and other wall-climbing animals in hopes of developing more large-scale, synthetic adhesives. For example, if you want to climb a smooth wall and can't make your sticky pads large enough to hold your body weight, there is an alternative: make your pads a lot stickier.

"We noticed that within closely related species pad size was not increasing fast enough to match body size, probably a result of evolutionary constraints. Yet these animals can still stick to walls," says Christofer Clemente from the University of the Sunshine Coast.

"Within frogs, we found that they have switched to this second option of making pads stickier rather than bigger. It's remarkable that we see two different evolutionary solutions to the problem of getting big and sticking to walls," says Clemente.

So, there may still be hope for Spiderman.

This study about gecko foot pads is published in the journal *Proceedings of the National Academy of Sciences*. Funding for this research was provided by the Biotechnology and Biological Sciences Research Council, Human Frontier Science Programme, Denman Baynes Senior Research Fellowship, and Discovery Early Career.

How Does a Solar Cell Create Electricity?

NGSS Standards: 3-PS2-3 Forces and Interactions

What It's About

Electricity impacts every part of our lives—how we cook food, how we wash our clothes, how we watch television or read a book late at night. Today, solar cells are being used more and more to use sunlight to create electricity to power our homes. How does electricity work? This experiment will demonstrate the basic idea of how tiny particles called electrons can move between objects, leading to an electrical force that can attract or push away other objects. This ability of electrons to move between objects leads to electricity, and has also led to fascinating technologies such as solar cells.

What You Need

- One or more balloons

- A small plate

- Several teaspoons of salt

- Several teaspoons of pepper

Useful Words

- *Electric charge*: a fundamental property of tiny particles of matter in every object around us. Positive and negative charges attract one another, and like charges repel one another.

- *Electricity*: the motion of electric charge that can be used to power motors, lights, and many other devices in our society

What to Do

1. Pour one teaspoon of salt and one teaspoon of pepper onto the small plate. Stir until thoroughly mixed.

2. Blow up one balloon, and then rub it hard against your hair. You should notice some of your hair will stand up straight if it's long enough!

3. Place the part of the balloon that you rubbed against your hair about 2.5 cm (1 inch) above the salt and pepper on the plate.

4. Write down your observations about what happens. Record roughly how much salt and how much pepper jump to the balloon.

5. Brush the salt and pepper attached to the balloon back onto the plate.

6. Then, rub the balloon against your hair again and place it about 5 cm (2 inches) above the plate.

7. Record your observations. Do you notice any difference between how much salt and pepper are transferred to the balloon?

8. Repeat the previous step while holding the balloon about 7.5 and 10 cm (3 and 4 inches) above the plate. Each time, record your observations.

9. What conclusions can you draw based on the data you gathered about how salt and pepper interact with the balloon?

Science to Know

When you rubbed the balloon against your hair, you created a lot of friction between the two surfaces. This friction knocks some electrons from your hair onto the balloon. Electrons are tiny particles with a negative electric charge that are inside every object in the universe! The balloon is now negatively charged because it has gained more electrons.

Now, when you place the balloon close to the salt and pepper without touching the plate, you should have seen individual pieces of salt and pepper suddenly jump to the balloon! Why does this happen? The negative charge on the balloon pushes away negatively charged electrons in the salt and pepper, because like charges repel. This only leaves tiny positively charged particles, called protons, in the salt and pepper that are still close to the balloon. Opposite charges attract, so this attractive force pulls the salt and pepper up to the balloon!

In the final steps of the experiment, you explored how changing the distance between the balloon and the plate affects how much salt and pepper jump to the balloon. You likely observed that as the balloon moved farther away, relatively more pepper than salt jumped to the balloon. This occurs because a single pepper flake is lighter than one piece of salt. The electrical attraction between the negatively charged balloon and the salt and pepper must overcome gravity to make the jump. As the balloon is moved farther away, only the lightweight pepper can overcome this downward gravitational pull to still reach the balloon.

This experiment demonstrated a simple way that anyone can observe electrical charge moving between objects. However, this same principle powers all the lights and appliances in your house through electricity! When you turn on a light or the dishwasher, electrons rush through the wires in your walls to provide the energy to run these devices.

To make these electrons move through the wires, we must give them energy. One way to do this is to use a solar cell that converts sunlight into a force to push electrons through the wires in your house. Solar cells separate and move electrical charge just like you did by rubbing the balloon against your hair, while solar cells achieve this using a complex combination of materials.

Application: Nanostructured Honeycomb Creates Electricity from Light

Zoom in to the nanometer scale—less than the width of a human hair—and you might think the new device designed by a team of scientists led by Lei Zhang is a honeycomb. Upon closer inspection, you would find that the hexagonal structure is made of gold and that a long string of organic molecules winds up and down through each hexagonal space. And one more thing: this device, so perfectly structured in the world of atoms and molecules, can create electricity from light.

These researchers from the Universities of Strasbourg and Nova Gorica have developed a novel structure that could overcome some of the obstacles faced by organic solar cells. As the name implies, these types of solar cells employ an organic molecule that absorbs light to create an excited and negatively charged electron. However, the electron leaves behind a positively charged hole when excited, and these oppositely charged particles need to be separated—electrons going to the cathode, holes to the anode—to generate current, for the solar cell to be useful. This separation process, which is very difficult to achieve in organic solar cells, is the focus of extensive scientific research.

One possible solution to this technical obstacle is to use an organic nanowire as a light absorber. The nanowire consists of a long chain of the organic absorbing material stacked one after another. The benefit of such a structure is its high surface-to-volume ratio, which means that electrons and holes spend little time close to one another before reaching the nanowire's surface and jumping their separate ways to the cathode and anode. This idea has tantalized researchers for some time; however, it has been difficult to create a structure that easily connects the nanowire to the electrodes.

Zhang's team has found a promising solution through a both practical and artful design. Beginning with a silicon substrate acting as the anode, the researchers patterned a gold honeycomb on top of it to act as the cathode. Finally, the organic nanowire, made of a molecule known as PTCDI-C8, was deposited to snake across the top of the honeycomb, then down along the exposed silicon base, then up along the gold honeycomb, and so forth, like a snake weaving through the hexagons. Such a unique design creates innumerable contacts between the organic wire and the gold (cathode) and silicon (anode), giving plenty of opportunities for the photoexcited electron and hole to separate and zoom away to create an electrical current.

The true beauty of this device may rest in its method of creation. Tiny—in fact, nanoscale—spheres were set onto the silicon substrate in a honeycomb shape to act as a mask, just like using tape to cover parts of a wall you don't want to paint. Then, gold was deposited over the entire silicon substrate. Finally, the spheres were etched away, leaving a perfect honeycomb gold electrode. This entire process creates a large enough honeycomb mesh to allow for hundreds of organic nanowires to lace through it.

Using this new nanowire device, half of all photoexcited electrons and holes could escape to the electrodes to contribute to the electrical current: a very good number for organic solar cells. In addition, Zhang and his team hope to use the device to test the fundamental properties of one-dimensional organic materials by varying the electrode size and measuring changes in performance. The device could also be used as a photodetector, and its design could be a template for light-emitting diodes. Above all, the finely tuned structure marks another significant advance in scientists' ability to control the patterns of nature at the levels of atoms and molecules.

How Is the Aurora Borealis Created?

NGSS Standards: 3-PS2-3 and 3-PS2-4 Forces and Interactions

What It's About

The aurora borealis is one of the most impressive natural phenomena we can see from the surface of Earth. Did you know that the aurora is created because of how particles from the sun interact with Earth's magnetic field? In this experiment, you will explore the fundamental properties of magnetic fields in your own home.

What You Need

- Iron filings (can be bought at most hardware stores)

- Two round neodymium or rare-earth magnets (can be bought at most hardware stores)

- Two pieces of white paper

- A pencil

Useful Words

- *Magnet*: an object with a north and south pole that has magnetic field lines connecting these two poles

- *Magnetic field:* a fundamental phenomenon of the universe that connects two opposite poles of a magnet. Other magnets will align along the direction of the magnetic field, and charged particles will rotate around the field.

- *Ferromagnet*: an object that has the special property of aligning with magnetic field lines

- *Magnetic force:* the push or pull on an object due to its interaction with a magnetic field

What to Do

1. Place the two magnets side-by-side underneath the piece of white paper, about 2.5–5 cm (1–2 inches) apart.

2. Slowly pour some iron filings onto the piece of paper on top of the location of the magnets. Observe and record what happens to the iron filings. Brush away any iron filings that are directly on top of the magnets.

3. Use the pencil to sketch the lines formed by the iron filings around the magnets. Do the lines connect the two magnets or move away from both?

4. Next, clean off all the iron filings. Then, flip one of the magnets over so it is still underneath the paper but its other surface is now touching the paper.

5. Slowly pour the iron filings onto the paper over the magnets again. Observe and record what happens to the iron filings. Have they changed the pattern they create around the magnets?

6. Use the pencil to sketch the lines formed by the iron filings. Then, compare this drawing to the first one. Compare and contrast the line patterns. What differences do you see?

Science to Know

Your drawings have uncovered an invisible phenomenon of nature that is around you all the time—magnetic fields! Magnetic fields cause a magnetic force that pushes or pulls magnets as well as particles with positive or negative charge; we don't always notice the magnetic force in our everyday life because the force of gravity from Earth is so much stronger.

In this experiment, you studied two scenarios: one with the two magnets in an initial configuration, and a second with one of the magnets flipped over. You should have noticed distinct differences in the patterns of your iron filings. These filings are *ferromagnets*, which means they are a very special type of material that will change their orientation to follow the invisible magnetic field lines created by the magnets. In this way, you used the filings to track the magnetic fields that you couldn't see with your eye.

In one configuration, the iron filings should have created lines connecting the two magnets. This pattern would indicate that one of the magnets had its north pole facing up against the paper and the other had its south pole facing up, because magnetic field lines connect north to south poles. The second configuration should have seen the magnetic field lines curve away from both magnets. This pattern would indicate that both magnets had either their north or south poles facing up, since field lines from the same pole will repel away from one another.

Application: How to Find an Aurora

The northern lights (aurora borealis) and southern lights (aurora australis) are fascinating scientifically. In fact, aurora is not unique to the Earth. We have observed aurora in the upper atmosphere of Jupiter and Saturn with various spacecraft and ground-based telescopes.

Solar Wind

The sun constantly emits streams of particles from its atmosphere out into the solar system. This emission is referred to as the solar wind. Sometimes there are solar storms or solar flares, resulting in heavier emissions than normal. If the Earth passes through one of these emissions, then the resulting auroras will be brighter than usual. If the sun is calm, then auroras may be dim or appear so far north or south that very few people see them.

Magnetic Fields and Rubber Bands

The Earth has a magnetic field surrounding it because of the iron-nickel core at the center of our planet. The magnetic field exiting from the core is responsible for the magnetic north and south poles we use when we navigate with a compass. It also creates a magnetic force field around the Earth, which extends into space.

As charged particles (electrons are negative, protons are positive) in the solar wind encounter the Earth's magnetic field, they travel along the field lines. On the sunward side, the field is compressed by the solar wind to be closer to the Earth; however, on the night side of the planet, the field stretches away from the planet like a tail. Eventually, the magnetic loops stretch so much that they break like an overstretched rubber band. A piece heads off into space away from Earth, while the other part snaps back toward Earth. The piece snapping back toward Earth accelerates the particles it captured into Earth's upper atmosphere.

When these particles hit molecules in Earth's atmosphere, they trigger light displays depending on the altitude and energy of the collision. Most of the molecules in Earth's atmosphere are either nitrogen or oxygen, so they are hit most frequently. Colors produced may be pink, red, yellow, green, blue, or violet. Occasionally, orange or white are produced. Typically, nitrogen will produce red, violet, or blue. Oxygen usually produces green or yellow. Reds generally are emitted above 240 km, greens at 100–240 km, purple and violet above 100 km, and blues at 80–100km.

Massive Electric Currents

The movement of charged particles in Earth's magnetic field produces powerful electric currents. In 1859, an aurora and the associated electrical storm were so powerful that people read newspapers at night by its light. Telegraph operations were disrupted as the current produced by the charged particles overwhelmed the normal currents used in the lines to transmit the signals. One pair of telegraph operators in Boston and Portland turned off their power and used the current created by the electrical storm to keep their transmissions going.

Global Planetary Index: Aurora Forecasting

Apps providing forecasts of probable auroras rely on scientific measurement and calculation of changes in the strength of Earth's magnetic field. Single measurements are tracked as a K-index. An aggregation of the measurement stations is used to produce a Planetary K-index (Kp). The higher the Kp value, the more likely an aurora will appear and the further south (or north in the southern hemisphere) it may display.

References

Abraham, J., Vasu, K. S., Williams, C. D., Gopinadhan, K., Su, Y., Cherian, C. T., . . . Nair, R. R. (2017). Tunable sieving of ions using graphene oxide membranes. *Nature Nanotechnology.* doi:10.1038/nnano.2017.21

Akbari, H. 2005. *Energy Saving Potentials and Air Quality Benefits of Urban Heat Island Mitigation (PDF)* (19 pp, 251K). Lawrence Berkeley National Laboratory.

Berdahl P. and S. Bretz. 1997. Preliminary survey of the solar reflectance of cool roofing materials. *Energy and Buildings 25:*149-158.

Bretz, S., Akbari, H., & Rosenfeld, A. (1998). Practical issues for using solar-reflective materials to mitigate urban heat islands. *Atmospheric Environment, 32(1),* 95-101. doi:10.1016/s1352-2310(97)00182-9

Centers for Disease Control and Prevention. (n.d.). *Climate change and extreme heat events.* Retrieved from https://www.cdc.gov/climateandhealth/pubs/ClimateChangeandExtremeHeatEvents.pdf

Center for Disease Control and Prevention. (2006). *Extreme Heat: A Prevention Guide to Promote Your Personal Health and Safety.*

Environmental Protection Agency. *Heat Island Impacts.* (2019, March 01). Retrieved from https://www.epa.gov/heat-islands/heat-island-impacts

Hu, A., Levis, S., Meehl, G., Han, W., Washington, W., Oleson, K., . . . Strand, W. (2015). Impact of solar panels on global climate. *Nature Climate Change, 6(3),* 290-294. doi:10.1038/nclimate2843

James, W. 2002. Green roads: research into permeable pavers. *Stormwater 3(2):*48-40.

Laaidi, K., Zeghnoun, A., Dousset, B., Bretin, P., Vandentorren, S., Giraudet, E., & Beaudeau, P. (2012). The impact of heat islands on mortality in Paris during the August 2003 heat wave. *Environmental health perspectives, 120(2),* 254–259. doi:10.1289/ehp.1103532

Lasar, M. (2012, May 2). *Ars Technica.* Retrieved from
https://arstechnica.com/science/2012/05/1859s-great-auroral-stormthe-week-the-sun-
touched-the-earth/

NGSS Lead States. 2013. *Next Generation Science Standards: For States, By States.*
Washington, DC: The National Academies Press.

NOAA / NWS Space Weather Prediction Center. (n.d.). Retrieved from
https://www.swpc.noaa.gov/

Staff, S. (2017, October 11). *Aurora borealis: What causes the northern lights and where
to see them.* Retrieved from https://www.space.com/15139-northern-lights-auroras-earth-
facts-sdcmp.html

United Nations, Department of Economic and Social Affairs, Population Division (2014).
World Urbanization Prospects: The 2014 Revision, Highlights (ST/ESA/SER.A/352).

Photo Credits

1. Ames fly-in, by Max Goldberg

2. Costa Rica, Puerto Viejo, Heredia, by Max Goldberg

3. Gehl-Mittelsted Farm, by Max Goldberg

4. Tulips at Reiman, by Max Goldberg

5. Exploring Northfield, by Max Goldberg

6. Gecko and ant, courtesy of A. Hackmann and D. Labonte

7. Solar panels on a roof, courtesy of the U.S. Department of Energy

8. Aurora Above Kirkjufell Mountain, by Steven Spence

About the Writers

Emily Rhode is a science communicator and municipal water resources educator. Her goal is to make science accessible and interesting for everyone. She has worked as an environmental educator, science teacher, and professional communicator and trainer with an M.Ed. in Science Education.

Steven Spence regularly contributes photography and general science articles to Science Connected. His extensive body of work in nature photography is both beautiful and informative. He enjoys photography, nature, and science, and is an avid fan of Calvin and Hobbes. Steven researches his subjects, showcases his concerns about the environment, and advocates for more biodiversity and fewer pesticides in agriculture.

Dr. Jonathan Trinastic earned his PhD in physics at the University of Florida and currently works in the utility industry using data science to create a cleaner and more resilient electrical grid. He is interested in renewable energy technology and sustainable energy policies.

About Science Connected

Science Connected is a 501(c)(3) nonprofit organization based in San Francisco, California. Our mission is to create affordable access to scientific research and STEM education for all learners.

At Science Connected, we bring together scientists and educators to provide learners with the scientific knowledge that will help employ them and empower them become the informed citizens, leaders, explorers, and innovators of tomorrow. But that's not all. We also work with researchers, journalists, universities, and industry leaders to provide cutting-edge research findings to people of all ages who continue to change our world for the better.

We develop exciting, research-based, supplemental resources for elementary, middle, and high school science education.

Visit us online a www.scienceconnected.org.

More from Science Connected

Think Globally, Garden Locally

This anthology is an Investigation into urban gardening, sustainable agriculture, and healthy pollinators, Learn about welcoming pollinators into your garden and growing plants without pesticides. Explore the relationship between chemicals and bee deaths and meet a scientist who became a beekeeper. Created by the Science Connected Magazine writers with financial support from our friends at the Clif Bar Family Foundation.

Purchase at https://amzn.to/2IwEQpm.

Earth Systems Lab Manual

This lab manual is full of science experiments for teachers and parents to enjoy with children ages 5-9 using items commonly found around the home. Explore the physical world. No specialized equipment is needed. These experiments are aligned with the Next Generation Science Standards (NGSS).

Purchase at https://amzn.to/2MMXAXc.

Science Connected Magazine

Science Connected journalists and scientists work together to bring you the latest scientific research in a format that everyone can read. We read peer-reviewed journal articles, ask questions, check facts, and write up the results for you.

Read for free at magazine.scienceconnected.org.

Discussion Guides

Bring Science Connected Magazine into your classroom with supplemental texts and discussion guides for middle school and high school.

Download them at scienceconnected.org/discussion-guides.